William Paterson Turnbull

The birds of east Pennsylvania and New Jersey

William Paterson Turnbull

The birds of east Pennsylvania and New Jersey

ISBN/EAN: 9783743388178

Manufactured in Europe, USA, Canada, Australia, Japa

Cover: Foto ©berggeist007 / pixelio.de

Manufactured and distributed by brebook publishing software
(www.brebook.com)

William Paterson Turnbull

The birds of east Pennsylvania and New Jersey

THE

BIRDS OF EAST PENNSYLVANIA AND NEW JERSEY.

BY

WILLIAM P. TURNBULL. LL.D.

AUTHOR OF THE "BIRDS OF EAST LOTHIAN;"
MEMBER OF THE ACADEMY OF NATURAL SCIENCES OF PHILADELPHIA;
OF THE LYCEUM OF NATURAL HISTORY, NEW YORK;
CORRESPONDING MEMBER OF THE NATURAL HISTORY SOCIETY OF GLASGOW, ETC.

———————————aves, solatia ruris,
Assuetum silvis, innocuumque genus,
Quæ facitis nidos, quæ plumis ova fovetis,
Et facili dulces editis ore modos.—*Ovid.*

PHILADELPHIA:

HENRY GRAMBO & CO., CHESTNUT STREET.

1869.

ALEXANDRI WILSONI,

ΤΟΥ ΟΡΝΙΘΟΓΝΩΜΟΝΟΣ,

NOMINIS MEMORLÆ,

NULLO IN ÆVO OBLIVISCENDÆ,

HOCCE PARVULUM,

AB EJUS MAXIMO OPERE DEDUCENS,

UT RIVULUS AB OCEANO,

PRIMORDIUM,

GULIELMUS P. TURNBULL.

D. D.

Ipse dedit Genitor, silvosa Columbia curat,
Arte suæ prolis Scotia pinxit, aves.

PREFACE.

In preparing the following Catalogue, the object of the writer has been to present in a simple and compact form the Ornithology of a small portion of North America, comprising that part of Pennsylvania eastward of the Alleghany Mountains, and of New Jersey, including the coast line which extends from Sandy Hook to Cape May. From the geographical position of the district it is particularly favourable for observation, being the resort, at some period of the year, of a large proportion of the birds of this continent; and, from the fact of its being the temporary resting-place of most of the migratory birds, there is probably no district of the same extent in this country that is frequented by such a number of species. A considerable number of our Summer visitants from the Gulf States and Mexico appear to make it their northern limit, while other flocks remain only a short period in Spring, and migrate still further north, penetrating as far as British America to breed; and these again arrive in Autumn on their return journey to their Winter retreats. It may likewise be noted that the district is the southern limit of many species which breed at Hudson's Bay and the fur countries, and pass the Winter on the Delaware and Chesapeake, thus forming a line of separation, so to speak, for the migratory flights of many interesting birds coming from opposite directions.

On consulting the list, however, it will be remarked that the proportion of what may be considered resident birds is small. This fact seems to have attracted observation as far back as the time of Dr Benj. S. Barton, who published a work entitled "Fragments of Natural History" about seventy years ago, in which it is stated that in the district now spoken of, very few species remained all the year, and that even of these there appeared to have been a partial migration in severe Winters; such birds, especially, as lived on insects and small fruits, being compelled to retire southwards—a fact still noticeable at the present day, many species that usually migrate remaining in mild and open Winters. Closer observation of late years, however, has enabled ornithologists to affirm with certainty that, of so-called migratory birds, a greater number pass the Winter with us than has been hitherto supposed. Of these, the Yellow-rump Warbler *(Dendroica coronata)* and the *Blue Bird (Sialia sialis)* may be cited as examples; and it is very probable that in suitable localities, especially in the southern and warmer counties, many more will yet be found.

The identity of some American species—chiefly water birds—with those of the Old World has long been a subject of anxious consideration among ornithologists, and with regard to a few of these there is still a diversity of opinion, a slight deviation in the size, and in the shade of the plumage being generally all that can be detected. In such cases a careful comparison of the habits, note, and nidification is of much importance, and cannot fail eventually to decide the question. In those instances where the birds are considered identical, the original scientific names are given; but where a decided difference in the size of an average number of specimens occurs, the distinction is noted. Of these, the Duck Hawk *(Falco anatum)* may be mentioned as an example, but even in this case it may be fairly questioned whether the disparity does not arise from a more bountiful supply and greater variety of food enjoyed by the American bird, combined with a larger extent of hunting ground than falls to the lot of its European congener, the Peregrine Falcon. It has been conclusively proved that many birds of the same species from different localties in this country, vary not only in size, but also in plumage, specimens from the Pacific coast being generally darker in colour than those from the interior; and Professor Baird, in a recently published paper on the subject, has suggested that this is possibly the result of greater exposure to the elements, and a want of such protection as the dense inland forests afford. He also states, that "while some Florida birds are characterised by larger bills than their more northern brethren, several of the birds of the middle and western provinces have an increase in the length of the tail, as compared with the same or allied species in the east;" and, as if in corroboration of these views, the characters of each are often found united, in intermediate specimens, near the boundary line of their respective districts.

In giving the dates of the arrival and departure of the various species enumerated, the writer wishes it to be understood that these are only mentioned as approximate, so much depending on the season being early or late, in illustration of which, it may be stated, that the wet and backward weather throughout last spring (1867) delayed the arrival of the Warblers fully two weeks. It should also be mentioned, that the abundance or scarcity of each species has reference, unless when otherwise noted, not to any special locality, but to the whole district represented.

The diffusion of well authenticated information regarding the distribution of American birds, is yet a matter for future observation. There have been, no doubt, many important contributions on this subject, of late years, yet the field is so extensive that many years must elapse before we can lay claim to a thorough knowledge of many important particulars, which patient research and well-timed energy alone can solve. The author of the present little work, while claiming for it the merit of careful observation, extending over a period of several years, at the same time believes the plan of the Catalogue to be capable of attaining more useful results, if enlarged in proportion to the nature of the districts investigated. It is, therefore, to be hoped that accurate observers may undertake similar records, by means of which the next great work on the ornithology of our country may contain a better collection of facts, representing the phenomena of the remoter districts, than has yet been obtained.

While making the usual acknowledgments to those who have so obligingly furnished the author with information, it gives him much pleasure to record his additional obligations to his friend, Mr Thomas S. Hutcheson, for his valuable services in superintending the work when passing through the press, and for other necessary aid, which, at this distance, is most thankfully appreciated. Nor must he omit to thank another friend and correspondent—Mr Robert Gray, Secretary to the Natural History Society of Glasgow —for useful notices regarding some of the birds of the remoter Hebrides of Scotland, between which district and some parts of our own continent there would appear to be an occasional inter-migration of species. On this subject, however, the author may at a future time, conjointly with that gentleman, bring all the ascertained facts and observations under the notice of ornithologists in a separate form.

W. P. T.

PHILADELPHIA, *January*, 1869.

NUMBER OF SPECIES.

S.	Summer visitants.	. . . .	. 114
W.	Winter visitants.	. . . .	. 57
S. & A.	Pass through in Spring and Autumn.	. . . .	. . 60
	Permanently resident.	. . . .	. 52
	Stragglers, or irregular visitants.	. . . . . .	. 59
			——
			342

THE

BIRDS OF EAST PENNSYLVANIA

AND NEW JERSEY.

ORDER—RAPTORES. (Rapacious Birds.)

Turkey Buzzard. . . *Cathartes aura.* TURKEY VULTURE.
Not uncommon. It is more fre-
quently seen in Summer, but many
remain during the Winter. It breeds
along the sea-coast of southern New
Jersey.

B

Golden Eagle. . . . *Aquila chrysaëtos.* RING - TAILED EAGLE. Very rare. A few are seen almost every Autumn. (S. & A.)

Bald Eagle. . . . . *Haliaëtus leucocephalus.* WHITE-HEADED EAGLE. Rather rare, and oftener seen in Spring, haunting the Delaware and larger streams inland. It breeds in New Jersey, on the sea-coast.

Fish Hawk. . . . *Pandion Carolinensis.* OSPREY. Not uncommon, especially about Great Egg Harbour, where it nestles in large communities. It arrives in the end of March, and departs early in October. Very closely allied to *Pandion haliaëtus* of the Old World—the difference being so slight as to make it almost a variety. (S.)

Duck Hawk. . . . . *Falco anatum.* GREAT FOOTED HAWK. Rare. During Autumn and Winter it frequents the marshes along the sea-coast and the courses of rivers, preying upon wild fowl. It breeds on the Alleghanies and the cliffs bordering the Susquehanna. This bird is very like the *Falco peregrinus* of Europe, but is larger, and is now believed to be a distinct species.

Pigeon Hawk. *Falco columbarius.* BULLET HAWK. A daring plunderer in poultry yards. It migrates in Spring to the north, where it breeds, returning in the Autumn. At this season, and also in Winter, it is not uncommon. (W.)

Sparrow Hawk. . . . *Falco sparverius.* This beautiful little Hawk is plentifully distributed.

Goshawk. . . . . . *Astur atricapillus.* Rare, arriving early in September from the North. Audubon mentions having found its nest in Pennsylvania. It is distinct from *A. palumbarius* of Europe. (W.)

Cooper's Hawk. . . . *Accipiter Cooperii.* Plentiful. It nestles on the mountain ridges of the Alleghanies.

Sharp-shinned Hawk. . *Accipiter fuscus.* SLATE-COLOURED HAWK. Abundant, building its nest on trees, but one was found, near Philadelphia, on the edge of a high rock.

Red-tailed Hawk. . . *Buteo borealis.* HEN HAWK. CHICKEN HAWK. Common. Much more frequent in Autumn and Winter, haunting meadows and cultivated districts.

Red-shouldered Hawk. *Buteo lineatus.* WINTER FALCON. Common, and especially along the sea-shore, but most abundant in Winter.

Broad-winged Hawk. . *Buteo Pennsylvanicus.* Rare. This Hawk is also more frequently seen in Winter.

Rough-legged Buzzard. *Archibuteo lagopus.* Not uncommon. May be seen coursing along rivers and marshes in Winter. (W.)

Black Hawk. . . . . *Archibuteo Sancti-Johannis.* Rather rare. It is generally found sailing at a low flight over the marshy flats of the Delaware and other large rivers. This is a northern species, but its nest has been once found in New Jersey. (W.)

Marsh Hawk. . . . *Circus Hudsonius.* MOUSE HAWK. HARRIER. Abundant on the salt marshes of the Jersey coast and on the Delaware. It seems more common in Winter, and is seldom met with in mature plumage. It is somewhat larger than *Circus cyaneus* of Europe, but closely allied.

Barn Owl. . . . . .   *Strix pratincola.* Not rare, and more
frequent in Spring and Autumn. Its
nest is generally found in a hollow
tree near marshy meadows. It is a
larger bird than *Strix flammea* of
Europe, and a distinct species.

Great Horned Owl. . .   *Bubo Virginianus.* CAT OWL. Rather
rare. It is found in the deep recesses
of swampy woods, where it breeds.
It has become, of late years, much
less plentiful, although frequently
seen in Winter.

Mottled Owl. . .   .   *Scops asio.* SCREECH OWL. Abundant.
The young, called the RED OWL, was
long considered a distinct species.

Long-eared Owl.   . .   *Otus Wilsonianus.* Rather rare. Its
nest has been occasionally found in
the woods near Philadelphia. It is
more frequently met with in Autumn
than at other seasons. This species
is very like *Otus vulgaris* of Europe,
but is rather darker in colour, and
is larger.

Short-eared Owl.   *Otus brachyotos.* MARSH OWL. Not
uncommon; arriving in November,
and departing in April It is seen
mostly on the meadows along the
Delaware and smaller streams. Mr
John Krider found its nest on Pecks
Beach, coast of New Jersey, in 1850;
and Audubon mentions having found
it breeding in the great Pine Swamp
of Pennsylvania. European specimens are somewhat lighter in colour,
but in habits they are alike. (W.)

Barred Owl. . . . .   *Syrnium nebulosum.* GREY OWL.
Common, but more abundant in
Winter. This Owl has been observed
frequently flying during the day.

Acadian Owl. . . . . *Nyctale acadica.* SAW WHET OWL.
LITTLE OWL. Rare. This handsome
little Owl is more frequently seen in
Winter, especially on the marshes of
the New Jersey coast.

Snowy Owl. . . . . *Nyctea nivea.* WHITE OWL. A rather
rare Winter visitant. (W.)

ORDER—SCANSORES. (Climbing Birds.)

Yellow-billed Cuckoo. *Coccygus Americanus.* RAIN CROW.
Common, arriving in the end of
April, and departing in the middle
of September. It is generally found
in thick woods and orchards. (S.)

Black-billed Cuckoo. *Coccygus erythrophthalmus.* Migrates about the same time as *Coccygus Americanus,* but is hardly so plentiful. It frequents the borders of small streams. Wilson first distinguished it from the preceding species. (S.)

Hairy Woodpecker. . *Picus villosus.* SPOTTED FLICKER. Not uncommon, and especially frequent in orchards. A larger variety is found in the northern counties of Pennsylvania, and has been called *Picus canadensis.*

Downy Woodpecker. *Picus pubescens.* SAPSUCKER. Plentiful. This species is very like *Picus villosus* in its markings, but is much smaller.

Red-cockaded Woodpecker. *Picus borealis.* Rare. A southern species, and migrating thence towards Winter. (S.)

Yellow-bellied Woodpecker. *Picus varius.* This is one of the most beautiful of our Woodpeckers, and is not uncommon. It is generally met with in Summer, arriving early in April, but a few remain during the Winter. (S.)

Great Black Woodpecker. *Picus pileatus.* PILEATED WOODPECKER. LOG-COCK. BLACK WOODCOCK. Not uncommon, but much more rare than formerly. It is more abundant towards the Alleghany Mountains.

Red-bellied Woodpecker. *Picus Carolinus.* Common, but more frequent in Summer; found mostly on the larger trees of the forest.

Red-headed Woodpecker. *Picus erythrocephalus.* Plentiful. Arriving in the latter part of April, and departing in September or beginning of October. It appears to be more numerous towards the mountains. (S.)

Golden-winged Wood-pecker.

Colaptes auratus. FLICKER. YELLOW-SHAFTED WOODPECKER. HIGH HOLE. Abundant, and generally distributed, but much more plentiful in Summer. There is a partial migration south in October, the birds returning about the end of March.

ORDER—VOLITORES. (Birds moving chiefly by Flight.)

Ruby-throated Humming Bird.

Trochilus colubris. HUMMING BIRD. Rather plentiful in Summer, and abundant in warm seasons, being met with from the end of April to the beginning of October. It is frequent in gardens, where its habit of roaming among the flowers has been so beautifully depicted by Wilson in the well-known verses, "When morning dawns." (S.)

Belted Kingfisher. . .

Ceryle alcyon. KINGFISHER. Not uncommon; arriving early in April, and departing in October. A few, however, remain during the Winter. It is more abundant inland than on the coast. (S.)

Whip-poor-Will. . . *Antrostomus vociferus.* Rather common, from the end of April to the
beginning of September. It appears
to be less frequent near the coast. (S.)

Night Hawk. . . . *Chordeiles popetue.* NIGHT JAR. BULL
BAT. Abundant from the middle of
April to September. Often seen high
in the air above the streets of Philadelphia, and its nest has frequently been
found on the roofs of warehouses. (S.)

Chimney Swallow. . . *Chaetura pelasgia.* Common. Arrives
in the middle of April, and departs
early in September. (S.)

ORDER—INSESSORES. (Perching Birds.)

Barn Swallow. . . . *Hirundo horreorum.* Common from
the end of March to the end of
September. (S.)

Cliff Swallow. . . . *Hirundo lunifrons.* A rather rare
Summer visitant, but every year
• increasing in numbers. It arrives in
April. (S.)

White-bellied Swallow. *Hirundo bicolor.* Rather plentiful. Comes late in March, and leaves early in September. (S.)

Sand Martin. . . . *Cotyle riparia.* BANK SWALLOW. Not uncommon on the high bank of a river or the sea-shore, arriving early in March, and leaving about the middle of October. (S.)

Rough-winged Swallow. *Cotyle serripennis.* This is a southern species, and not rare. It arrives early in April, and seems to be more abundant in the lower counties. (S.)

Purple Martin. . . . *Progne subis.* Abundant, coming early in April, and leaving in the end of August. (S.)

Tyrant Flycatcher. . *Tyrannus Carolinensis.* KING BIRD. FIELD MARTIN. BEE BIRD. Plentiful. It arrives about the end of April, and leaves in September. (S.)

Great Crested Flycatcher. *Myiarchus crinitus.* Not uncommon, appearing early in May, and leaving in the end of September. (S.)

Pewee Flycatcher. . . *Sayornis fuscus.* PEWEE. PHŒBE BIRD. Rather plentiful, especially on the borders of creeks and small streams. It appears early in March, and leaves in November. (S.)

Olive-sided Flycatcher. *Contopus borealis.* Very rare. It is generally seen early in May on its way north, and returns in September. (S. & A.)

Wood Pewee. . . . *Contopus virens.* Not uncommon from the beginning of May to the middle of September. This species closely resembles the PHŒBE BIRD. (S.)

Traill's Flycatcher. . *Empidonax Traillii.* Rare, but some seasons it is not uncommon in the Spring, arriving about the middle of May. (S. & A.)

c

Least Flycatcher. . .   *Empidonax minimus.* Rather rare,
arriving in April on its northern
migration, and returning early in
September. A few remain to breed.
(S. & A.)

Green-crested Flycatcher. *Empidonax acadicus.* SMALL PEWEE.
Frequent from the beginning of May
to the middle of September. It is
generally found in the most secluded
parts of woods. (S.)

Yellow - bellied Fly-   *Empidonax flaviventris.* Rare. It
catcher.   arrives in the middle of April
on its way north. Dr Slack
found it breeding near Trenton.
(S. & A.)

Wood Thrush. . . .   *Turdus mustelinus.* SONG THRUSH.
One of our sweetest songsters, and
plentifully distributed. It arrives
about the middle of April, and
departs in October. (S.)

Hermit Thrush. . .   *Turdus Pallasii.* Not uncommon. It
arrives in April, and again late in
October, on its way south, when
it is more abundant along the sea-
coast. This bird is very like
Turdus Swainsonii, but the tail and
tail coverts are rufous. A few have
been observed during Winter when
that season has been open and mild.
(S. & A.)

Wilson's Thrush. .   *Turdus fuscescens.* TAWNY THRUSH.
Plentiful from the beginning of April
to October—a few remaining during
Winter. Many migrate further north
to breed. (S.)

Olive-backed Thrush.   *Turdus Swainsonii.* Rare. It arrives
in the end of April. A few also
of this species have been seen in
Winter. (S. & A.)

Grey-cheeked Thrush.   *Turdus aliciae.* Very rare. It has lately been found in the southern counties of Pennsylvania, and is not very uncommon in Maryland. (S.)

Migratory Thrush. . *Turdus migratorius.* ROBIN. Very abundant. During Autumn and Winter it migrates in flocks.

Mocking Bird. . . . *Mimus polyglottus.* This far-famed songster is now rare, arriving late in April, but it appears to have been plentiful in former years; and, according to Bartram and Dr Barton, even remained all the Winter near Philadelphia. (S.)

Cat Bird. . . . *Mimus Carolinensis.* CATTIE. Abundant from the middle of April to late in October. (S.)

Brown Thrush. . . *Mimus rufus.* THRASHER. FRENCH MOCKING BIRD. Common. Arrives early in April, and departs in October, but a few remain during mild Winters. This is one of the finest of our song birds. (S.)

Blue Bird. . . . *Sialia sialis.* RED-BREASTED BLUE BIRD. Abundant, and may be seen as early as the latter part of February. It leaves late in November, but so many remain during Winter, that it may be called a resident species.

Ruby-crowned Wren. *Regulus calendula.* Plentiful, arriving early in April, and returning from the north late in September. (S. & A.)

Golden-crested Wren. *Regulus satrapa.* Not uncommon. It comes in April, and again in September, a few remaining during the Winter. It is distinct from the *Regulus cristatus* of Europe. (S. & A.)

Tit-Lark. . . *Anthus Ludovicianus*. PIPIT. BROWN LARK. Common. It arrives from the north in October, and departs in April, but is more frequent in the Autumn and Spring migrations. (W.)

Creeping Warbler. . *Mniotilta varia*. BLACK AND WHITE CREEPER. Plentiful from the middle of April to the beginning of October. (S.)

Blue Yellow - backed Warbler. *Parula Americana*. BLUE YELLOW-BACK. Common from the end of April to October. (S.)

Maryland Yellow-throat. *Geothlypis trichas*. Abundant from the end of April to the beginning of September. (S.)

Mourning Warbler. . *Geothlypis Philadelphia*. Rare, and excessively so in Autumn. It arrives in the middle of May on its way north. (S. & A.)

Connecticut Warbler. *Oporornis agilis*. Rather rare. This species is very seldom met with in Spring, but is, however, more frequent in Autumn, appearing late in August. (S. & A.)

Kentucky Warbler. . *Oporornis formosus*. A southern species, and rather scarce. It arrives late in April. (S.)

Yellow-breasted Chat. *Icteria virens*. Common, coming early in May and leaving in September. (S.)

Worm-eating Warbler. *Helmitherus vermivorus*. Rather rare, arriving in the middle of May, and departing in the end of September. (S.)

Blue-winged Yellow Warbler. *Helminthophaga solitaria*. Somewhat rare, arriving in the middle of May. (S.)

Golden-winged Warbler. *Helminthophaga chrysoptera*. Rather rare. It appears about the end of April. (S. & A.)

Nashville Warbler. . *Helminthophaga ruficapilla.* Frequent. It arrives late in April, and is more plentiful in the spring and autumnal migrations. (S.)

Tennessee Warbler. *Helminthophaga peregrina.* Rather rare, but sometimes not uncommon in September on its return south. (S. & A.)

Golden-crowned Thrush. *Seiurus aurocapillus.* OVEN BIRD. Common from the end of April to late in September. (S.)

Water-Thrush. . . . *Seiurus noveboracensis.* Not uncommon, arriving in the end of April, and again late in August. (S. & A.)

Large - billed Water Thrush. *Seiurus Ludovicianus.* Not rare, but more abundant in the lower counties, and said to be plentiful round Washington. (S.)

Black-throated Green Warbler. *Dendroica virens.* Rather frequent, arriving early in May, and departing in October. It is more plentiful in the Spring and Autumn migrations. (S.)

Black - throated Blue Warbler. *Dendroica cœrulescens.* Abundant. It comes late in April, and again towards the end of October. A few remain to breed. (S. &. A.)

Yellow-rump Warbler. *Dendroica coronata.* MYRTLE BIRD. GOLDEN-CROWNED WARBLER. Common, appearing early in April, and again in October. It has been seen frequently during winter. (S. & A.)

Blackburnian Warbler. *Dendroica Blackburniæ.* Not uncommon. This beautiful warbler arrives early in May, returning in September. Its nest has been frequently found in Pennsylvania. The male in Autumn has been called HEMLOCK WARBLER. (S. & A.)

Bay-breasted Warbler.   *Dendroica castanea.*   Rather rare,
appearing late in April, and again
in October.   The young bird of this
species is the AUTUMNAL WARBLER
of Wilson.   (S. & A.)

Pine-creeping Warbler.   *Dendroica pinus.*   This species is
rather rare in Spring, but plentiful
in Autumn.   Arriving early in April,
and again in October.   Many remain during summer.   (S.)

Chestnut-sided Warbler. *Dendroica Pennsylvanica.*   Not uncommon.   Arriving early in May.
A few remain to breed.   (S. & A.)

Blue Warbler.   .   .   .   *Dendroica cœrulea.*   Rare.   From
May to end of August.   The BLUE
MOUNTAIN WARBLER of Wilson is
the young of this species.   (S.)

Black-Poll Warbler.   .   *Dendroica striata.*   Common.   It
arrives late in April, and again in
September.   (S. & A.)

Yellow Warbler.   .   .   *Dendroica œstiva.*   SUMMER YELLOW
BIRD.   Abundant.   ·Coming late in
April, and departing in the end of
September.   (S.)

Black and Yellow   *Dendroica maculosa.*   MAGNOLIA
   Warbler.   WARBLER.   One of our most beautiful warblers, and rather frequent,
coming early in May, and again in
September.   (S. & A.)

Cape May Warbler.   .   *Dendroica tigrina.*   Very rare; it
arrives early in May, and again
visits us on its way south about the
10th of October.   (S. & A.)

Yellow Red-Poll Warbler. *Dendroica palmarum.*   Plentiful.
Appearing early in April, and again
in September.   (S. & A.)

Prairie Warbler.   .   .   .   *Dendroica discolor.*   Not uncommon,
from the beginning of May to
October.   (S.)

Hooded Warbler. . . *Myiodioctes mitratus.* Rather rare; it arrives in April. (S.)

Green Black-cap Fly-catcher. *Myiodioctes pusillus.* This bird appears early in May, and again in October, and is rather abundant. (S. & A.)

Canada Flycatcher. . *Myiodioctes Canadensis.* Frequent, from the end of April to October. (S.)

Redstart. . . . . . *Setophaga ruticilla.* Abundant, from the end of April till late in September. (S.)

Scarlet Tanager. . . *Pyranga rubra.* FIRE BIRD. Plentiful, coming early in May, and leaving in September. (S.)

Summer Red Bird. . . *Pyranga œstiva.* RED BIRD. Rather rare, and found mostly in the southern counties of New Jersey, from May to the middle of August. (S.)

Cedar Bird. . . . *Ampelis cedrorum.* PRIB CHATTERER. Abundant, but less frequent in winter than at other seasons.

Great Northern Shrike. *Lanius borealis.* BUTCHER BIRD. Not uncommon, especially in winter. In March it migrates to the north, but many nestle on the mountain ridges of the Alleghanies.

Red-eyed Flycatcher. . *Vireo olivaceus.* Plentiful from the beginning of May to the middle of October. (S.)

Philadelphia Vireo. . *Vireo Philadelphicus.* SHORT-BILLED VIREO. Very rare. (S. & A.)

Warbling Flycatcher. *Vireo gilvus.* Frequent from early in May to the middle of October. (S.)

White-eyed Vireo. . . *Vireo noveboracensis.* Common from early in April to October. (S.)

Blue-headed Flycatcher. *Vireo solitarius.* SOLITARY VIREO. Rather rare, arriving in April, and departing in October. (S.)

Yellow-throated Fly-catcher.	*Vireo flavifrons.* Not uncommon from the end of April to September. (S.)
Great Carolina Wren.	*Thryothorus Ludovicianus.* Rather rare. It appears early in May on the borders of the Delaware. Mr John Cassin informs me that he has occasionally seen this bird on the Wissahickon in Winter. (S.)
Bewick's Wren. . . .	*Thryothorus Bewickii.* Rare. It arrives early in May. (S.)
Long-billed Marsh Wren.	*Thryothorus palustris.* Common along the Delaware and other streams from the beginning of April to the middle of October. (S.)
Short-billed Marsh Wren.	*Thryothorus stellaris.* Rather rare from April to September. (S.)
House Wren.	*Troglodytes aedon.* Abundant from the end of April to October. (S.)
Wood Wren.	*Troglodytes Americanus.* Rare. This bird very closely resembles the preceding species, *Troglodytes aedon,* in Autumn plumage; indeed, it is doubtful if the two are distinct. (S.)
Winter Wren. . . .	*Troglodytes hyemalis.* Not uncommon, arriving in October. Its nest has been occasionally found in Pennsylvania. This species is so very like *Troglodytes vulgaris* of Europe as to be at least only a variety. (W.)
Tree Creeper. . . .	*Certhia familiaris.* GREY CREEPER. Abundant, but more so in winter than at other times of the year. I can detect no difference between it and that of Europe; it has the same shrill but feeble note, and its habits are identical.
White-bellied Nuthatch.	*Sitta Carolinensis.* Plentiful, and resident.

Red-bellied Nuthatch. . *Sitta Canadensis.* Rather rare, from October to April. (W.)

Blue-grey Gnatcatcher. *Polioptila cærulea.* BLUE-GREY FLY-CATCHER. Not uncommon from the middle of April to the end of September. (S.)

Tufted Titmouse. . . *Lophophanes bicolor.* TOM-TIT. Very common, and especially abundant in Summer.

Black-cap Titmouse. . *Parus atricapillus.* CHICKADEE. Abundant.

Carolina Titmouse. . . *Parus Carolinensis.* SMALL BLACK-CAP TITMOUSE. A southern species, and rather rare. (S.)

Shore Lark. . . . . *Alauda alpestris.* SKY LARK. HORNED LARK. Plentiful, appearing late in October, and generally leaving in March, but some seasons it remains until April. (W.)

Pine Grosbeak. . . . *Pinicola Canadensis.* PINE BULL-FINCH. Rather rare. This species is very like *Pinicola enucleator* of Europe, but is larger. It may be a variety only. (W.)

Purple Finch. *Carpodacus purpureus.* Not uncommon from September to April, a few remaining during Summer. (W.)

American Goldfinch. . *Chrysomitris tristis.* YELLOW BIRD. THISTLE BIRD. Common, but is less frequent in Winter than at other times.

Pine Finch. . . . . *Chrysomitris pinus.* PINE LINNET. Plentiful. Arriving late in October, and departing in March. (W.)

Red Crossbill. . . . *Loxia Americana.* COMMON CROSS-BILL. Not uncommon in winter, and resident in the Alleghanies. It nearly resembles *Loxia curvirostra* of Europe, but is smaller.

D

White-winged Crossbill. *Loxia leucoptera.* Rare, but in some winters more plentiful. Large flocks have been known to cross the Atlantic to Britain, and they are occasionally seen to alight on vessels at sea. See Gray's "Birds of the West of Scotland." (W.)

Lesser Red-Poll. . . . *Linota linaria.* HEMP BIRD. Not uncommon in severe winters, coming early in November, and remaining until April. (W.)

Snow Bunting. . . . *Plectrophanes nivalis.* WHITE SNOW BIRD. Of frequent occurrence, and usually appearing after a snow storm. It arrives early in December, and leaves in March. (W.)

Savannah Sparrow. . . *Passerculus Savanna.* Common, being frequent in Summer near the mountains, and plentiful in Winter on the sea shore.

Grass Finch. . . . . *Pooecetes gramineus.* GRASS SPARROW. BAY-WINGED SPARROW. Abundant, its numbers being augmented in Summer by flocks arriving early in April.

Yellow-winged Sparrow. *Coturniculus passerinus.* Common, arriving late in April, and departing in October. (S.)

Henslow's Bunting. . . *Coturniculus Henslowi.* Rather rare. (S.)

Sharp-tailed Finch. . . *Ammodromus caudacutus.* QUAIL HEAD. Frequent on the salt marshes along the coast. (S.)

Sea-side Finch. . . . *Ammodromus maritimus.* GREY SHORE FINCH. Abundant on the sea shore. (S.)

White-crowned Sparrow. *Zonotrichia leucophrys.* CHIP BIRD. Rather rare; from September to the beginning of May. (W.)

White-throated Sparrow. *Zonotrichia albicollis.* Common, appearing early in October, and leaving in April. My friend, Mr Gray of Glasgow, has informed me that this familiar bird has been killed in Aberdeenshire, Scotland. Female specimens in Autumn plumage bear, indeed, a somewhat near resemblance to one or two Buntings, for which they may be mistaken; so that it is not unlikely it may be found again if closely looked for. (W.)

Snow Bird. . . . *Junco hyemalis.* BLACK SNOW-BIRD. CHUCK-BIRD. Abundant in Winter from October to April. A few remain to breed on the Alleghanies. (W.)

Tree Sparrow. . . . *Spizella monticola.* CANADA BUNT-ING. Plentiful from the end of October to April. (W.)

Field Sparrow. . . . *Spizella pusilla.* RUSH SPARROW. Common. Comes early in April, and leaves in October. (S.)

Chipping Sparrow. . . *Spizella socialis.* CHIPPY. Abundant from the end of March to October. (S.)

Song Sparrow. . . . *Melospiza melodia.* Abundant from the beginning of March to November, but many remain all Winter.

Lincoln's Finch. . . *Melospiza Lincolnii.* Rather rare. It appears early in May, and again in Autumn. (S. & A.)

Swamp Sparrow. . . *Melospiza palustris.* RED GRASS-BIRD. MARSH SHORE FINCH. Common. Arriving about the middle of April. (S.)

Fox-coloured Sparrow. *Passerella iliaca.* FOX SPARROW. Frequent from the end of October to March. (W.)

House Sparrow. . . *Passer domesticus.* According to Mr G. N. Lawrence, this well-known European bird has been successfully introduced into Jersey City, Hoboken, and Newark, N.J., where they excite much interest, and have become great favourites. A very interesting account of the species in its new haunts, is given by that gentleman, on the Birds of New York, in the "Annals of the Lyceum of Natural History of New York," Vol. VIII., April, 1866.

Black-throated Bunting. *Euspiza Americana.* Plentiful. Appearing early in May, and leaving in September. (S.)

Rose-breasted Grosbeak. *Guiraca Ludoviciana.* Rather rare. It comes in April, and leaves early in September. (S.)

Indigo Bird. .   . . *Cyanospiza cyanea.* Not uncommon from May to September. (S.)

Cardinal Grosbeak. . . *Cardinalis Virginianus.* RED BIRD. VIRGINIA NIGHTINGALE. Frequently met with. Many remain throughout the Winter in the lower counties.

Ground Robin.   . *Pipilo erythrophthalmus.* TOWHEE. CHEEWINK. Abundant from the middle of April to October. A few remain during Winter, and may be found in well-sheltered localities. (S.)

Reed Bird.   . . . . *Dolichonyx oryzivorus.* RICE BUNTING. BOB-O-LINK. Plentiful. It appears early in May, and again in the middle of August, when it arrives from the north in large numbers, spreading along the Delaware and other rivers. (S. & A.)

Cow Bird. . . . . . *Molothrus pecoris.* COW BLACKBIRD. COW-PEN BIRD. Common. It comes late in March, and disappears in October. Like the Cuckoo of Europe, it deposits its eggs in the nests of other birds. (S.)

Red-winged Blackbird. *Agelaius phœniceus.* BLACKBIRD. RED-WINGED STARLING. SWAMP BLACKBIRD. Abundant from the beginning of March to the end of October, appearing in Autumn in very large flocks. (S.)

Meadow Lark. . . *Sturnella magna.* OLD FIELD LARK. Plentiful. There is a partial migration southwards in severe Winters.

Orchard Oriole. . . . *Icterus spurius.* Common. Appearing early in May, and leaving in September. (S.)

Baltimore Oriole. . . *Icterus Baltimore.* GOLDEN ROBIN. HANG NEST. Frequent, coming late in April, and disappearing early in September. (S.)

Rusty Grakle. . . . *Scolecophagus ferrugineus.* RUSTY BLACKBIRD. Not uncommon. It arrives late in March, and again in October. (S. & A.)

Crow Blackbird. . . . *Quiscalus versicolor.* PURPLE GRAKLE. Plentiful, arriving early in March, and leaving in November. (S.)

Raven. . . . . . . *Corvus carnivorus.* Rare. A few are found on the Jersey coast, and in the direction of the Mountains. This bird is so like the Common Raven of Europe (*Corvus Corax*) as to be merely a variety.

Common Crow. . . . *Corvus Americanus.* Abundant. This species is very like *Corvus corone* of Europe, but entirely different in habits, congregating in great flocks.

Fish Crow. . . . . *Corvus ossifragus.* Not uncommon
along the sea-shore, and on the
Delaware. It arrives early in
April. (S.)

Blue Jay. . . . . . *Cyanura cristata.* Frequent, but less
numerous in Winter than at other
seasons.

ORDER—RASORES. (Gallinaceous Birds.)

Wild Pigeon. . . . . *Ectopistes migratoria.* PASSENGER
PIGEON. Plentiful, but is more
frequent in Spring and Autumn,
when it congregates in large flocks.

Common Dove. . . . *Ectopistes Carolinensis.* CAROLINA
TURTLE DOVE. Common from
March to October, but many remain
during the Winter.

Wild Turkey. . . . *Meleagris gallopavo.* Now rare. A
few straggling flocks are yet met
with on the Alleghanies, and specimens which have been killed there
may be seen every Winter in the
Philadelphia market. It is not
uncommon in Virginia.

Pinnated Grouse. . . *Cupidonia cupido.* PRAIRIE HEN. PRAIRIE CHICKEN. Now very rare. A few are still met with in Monroe and Northampton counties, Pennsylvania, where I have shot the species. Within the last year or two it has also been found on the Jersey Plains.

Ruffed Grouse. . . . *Bonasa umbellus.* PHEASANT. Plentiful, and more so in the direction of the Alleghany mountains.

Partridge. . . . . . *Ortyx Virginianus.* QUAIL. BOB-WHITE. Abundant, and plentifully distributed, but less common than formerly, near the larger cities.

ORDER—GRALLATORES. (Wading Birds.)

Snowy Heron. . . . *Garzetta candidissima.* WHITE-CRESTED EGRET. WHITE POKE. Not uncommon on the salt marshes of the sea-coast, from the beginning of April to October. This species is quite distinct from *Garzetta egretta* of the old world. (S.)

White Heron. . . . *Ardea egretta.* WHITE CRANE. GREAT WHITE EGRET. Rather rare, arriving about the middle of May. It is distinct from *Ardea alba* of Europe. (S.)

Great Blue Heron. . . *Ardea herodias.* BLUE CRANE. Common, arriving in April. A few, however, remain during Winter. (S.)

Blue Heron. . . . . *Ardea caerulea.* Rare, but it has been found breeding at Cape May. (S.)

Least Bittern. . . . *Ardetta exilis.* Not uncommon along the Delaware. It arrives early in May. (S.)

Bittern. . . . . . *Botaurus lentiginosus.* QUAWK. DUNKADOO. INDIAN HEN. Plentiful from the middle of April to October. (S.)

Green Heron. . . . *Butorides virescens.* FLY-UP-THE-CREEK. Abundant from early in April to October. (S.)

Night Heron. . . . *Nyctiardea Gardeni.* QUA BIRD. Frequent. Arriving early in April, and leaving in October. This species is very like *Nyctiardea grisea* of Europe, but is, however, distinct. (S.)

Golden Plover. . . . *Charadrius Virginicus.* FROST BIRD. BULL HEAD. Common, appearing in the end of April, and again early in September. It is smaller than *Charadrius pluvialis* of Europe, and quite distinct from that species. (S. & A.)

Killdeer Plover. . . . *Aegialitis vociferus.* KILLDEER. Plentiful, and especially abundant along the sea-shore in Winter.

Wilson's Plover. . *Aegialitis Wilsonius.* Rather rare. Arrives early in May. On the original drawing of this bird now before me, by Wilson himself, I find he has named it GREAT-BILLED PLOVER. (S.)

Semipalmated Plover. . *Aegialitis semipalmatus.* RING PLOVER. RING NECK. Frequent on the sea-coast, appearing late in April, and again in September. (S. & A.)

Piping Plover. . . . *Aegialitis melodus.* BEACH BIRD. Common on the sea-shore from the end of April to October. (S.)

Grey Plover. . . . . *Squatarola cinerea.* BLACK-BELLIED PLOVER. WHISTLING FIELD PLOVER. BEETLE - HEADED PLOVER. Plentiful. It appears late in April, and again in September, a few remaining on the uplands to breed. (S. & A.)

Oyster Catcher. . . . *Hœmatopus palliatus.* FLOOD GULL. Rather scarce. This bird is very like *Hœmatopus ostralegus* of Europe, and may be only a variety. (S.)

Turnstone. . . . . *Strepsilas interpres.* CALICO BACK. BRANT BIRD. Also called HORSEFOOT SNIPE, from its feeding on the spawn of the King Crab. Abundant, arriving early in April, and returning south in October. (S. & A.)

Avocet. . . . . . *Recurvirostra Americana.* BLUESTOCKING. Rather rare, appearing late in April, and leaving in October. (S.)

Black-necked Stilt. . *Himantopus nigricollis.* THE LAWYER. LONGSHANKS. Rather scarce, from the end of April to September. I have found its nest on Egg Island, Delaware Bay. (S.)

Northern Phalarope. . *Phalaropus hyperboreus.* LOBEFOOT. This species, the RED - NECKED PHALAROPE of British ornithologists, is rare, arriving early in

E

May, and again in September, being, however, more frequent in Autumn. (S. & A.)

Woodcock. . . . . *Philohela minor.* Plentiful, from early in March until November, a few remaining during Winter. (S.)

Snipe. . . . . . . *Gallinago Wilsonii.* ENGLISH SNIPE. WILSON'S SNIPE. Abundant, arriving early in March, and again in September. A few remain during Summer. It is very like *Scolopax gallinago* of Europe, but is distinct, and also differs somewhat from that species in its habits. (S. & A.)

Red-breasted Snipe. . *Macrorhamphus griseus.* GREY SNIPE. BROWN BACK. QUAIL SNIPE. DOWITCHER. Not uncommon, appearing early in April, and again in August. (S. & A.)

Knot. . . . . . . *Tringa canutus.* RED-BREASTED SANDPIPER. ROBIN SNIPE. ASH-COLOURED SANDPIPER. GREY BACK, and also frequently called WHITE ROBIN SNIPE, in its Autumn plumage. Common. It arrives in May, on its way north; returning about the middle of August. (S. & A.)

Purple Sandpiper. . . *Tringa maritima.* Very rarely seen so far south, and generally in Winter. (W.)

Red-backed Sandpiper. *Tringa alpina, var.: Americana.* DUNLIN. BLACK-BREAST, and in Autumn, WINTER SNIPE. Abundant. It arrives in April, and again in September, a few remaining during Winter. American specimens are larger, and have the bill somewhat longer than those of Europe. The difference, indeed, is greater than between the true *Tringa alpina* and *Tringa Schinzii.* (S. & A.)

Pectoral Sandpiper. . . *Tringa maculata.* JACK SNIPE. MEADOW SNIPE. Plentiful, arriving in April, and again appearing about the end of August, when they are most abundant. (S. & A.)

Least Sandpiper. . . *Tringa Wilsonii.* WILSON'S SANDPIPER. PEEP. OXEYE. Abundant in the early part of May, and again in August. I once saw a flock on Brigantine Beach as early as 20th July. (S. & A.)

Bonaparte's Sandpiper. *Tringa Bonapartii.* LITTLE SNIPE. Frequent, and oftener met with on its return south in Autumn. This bird was long confounded with *T. Schinzii* of Europe. (S. & A.)

Sanderling. . . . . *Calidris arenaria.* SANDERLING PLOVER. RUDDY PLOVER. Abundant on the sea-coast, arriving in May, and again in the end of August, but many remain throughout the Winter. (S. & A.)

Semipalmated Sandpiper. *Ereunetes petrificatus.* PEEP. Plentiful on the coast early in May, returning from the north in August and September. (S. & A.)

Stilt Sandpiper. . . *Micropalama himantopus.* LONG-LEGGED SANDPIPER. Very rare. Seen in May, and again in August. (S. & A.)

Willet. . . . . . . *Symphemia semipalmata.* SEMIPALMATED TATLER. STONE CURLEW. Common from the middle of April to October. (S.)

Tell-Tale. . . . . . *Gambetta melanoleuca.* TELL-TALE GODWIT. VARIED TATLER. GREATER YELLOW-SHANKS. Plen-

	tiful, from the middle of April until November. (S.)
Yellow-shanks Tatler.	*Gambetta flavipes.* YELLOW-LEGGED SNIPE. Abundant. Appears late in April, and again in the end of August. Many remain during the Summer. (S. & A.)
Solitary Sandpiper. .	*Rhyacophilus solitarius.* WOOD TAT-LER. Not uncommon from May to September. (S.)
Spotted Sandpiper. .	*Tringoides macularius.* TILTUP. PEET-WEET. SPOTTED SAND LARK. Abundant from the beginning of April to the end of October. (S.)
Bartram's Sandpiper. .	*Actiturus Bartramius.* FIELD PLO-VER. UPLAND PLOVER. GRASS PLOVER. Plentiful from the middle of April till late in September. (S.)
Buff-breasted Sandpiper.	*Tryngites rufescens.* Rather rare, and is generally seen late in Autumn. (S. & A.)
Great Marbled Godwit.	*Limosa fedoa.* MARLIN. Not un-common. It arrives in May, and returns from the north in the end of September. (S. & A.)
Hudsonian Godwit. .	*Limosa Hudsonica.* RING - TAILED MARLIN. Rather scarce. It arrives late in September. (S. & A.)
Long-billed Curlew. .	*Numenius longirostris.* SICKLE BILL. BIG CURLEW. Frequent, arriv-ing early in May, and again in September. (S. & A.)
Hudsonian Curlew. .	*Numenius Hudsonicus.* SHORT-BILLED CURLEW. JACK CURLEW. Plentiful, arriving on its way north in May, and returning about the end of August. This species is not unlike *Numenius phœopus* of Europe. (S. & A.)

Esquimaux Curlew, . *Numenius borealis.* LITTLE CURLEW. Rather rare, appearing in May, and again in September. (S. & A.)

King Rail. . . . . . *Rallus elegans.* MARSH HEN. FRESH-WATER MUD HEN. This large and handsome Rail is rather scarce. (S.)

Clapper Rail. . . . . *Rallus crepitans.* MEADOW HEN. MUD HEN. Abundant on the salt marshes along the sea-coast, from the middle of April to late in September. (S.)

Virginia Rail. . . . *Rallus Virginianus.* FRESH-WATER MARSH HEN. Not uncommon along the Delaware and other streams, arriving late in April, and leaving late in October. (S.)

Common Rail. . . . *Porzana Carolina.* RAIL. SORA. CAROLINA RAIL. Abundant, arriving from the south early in May. About the beginning of August it returns from the north in great numbers, and finally leaves us in October. A few remain to breed during Summer. (S. & A.)

Little Black Rail. . . *Porzana Jamaicensis.* LEAST WATER RAIL. Rare. It breeds on the marshes of Cape May county, New Jersey. (S.)

Yellow Rail. . . . . *Porzana Noveboracensis.* LITTLE YELLOW RAIL. YELLOW-BREASTED RAIL. Rare, coming about the end of April, and leaving late in October. (S.)

Coot. . . . . . . *Fulica Americana.* HEN BILL. WHITE BILL. Rather rare. It appears early in April, remaining till November. It is distinct from *Fulica atra* of Europe. (S.)

Florida Gallinule. . .   *Gallinula galeata.*   WATER HEN.
COMMON GALLINULE. A very rare
Summer visitant, from the middle
of May to late in October, on the
Delaware and Susquehanna. It is
very like *Gallinula chloropus* of
Europe, but is distinct. Mr John
Cassin has informed me that one
of these birds flew into a house in
Philadelphia last season, by an open
window, and that he kept it for
some time in confinement. (S.)

ORDER—NATATORES.   (Swimming Birds.)

American Swan. . . .   *Cygnus Americanus.* Frequently seen
on its Spring and Autumn migra-
tions.   It winters on Chésapeake
Bay.   (W.)

Snow Goose. . . . .   *Anser hyperboreus.*   WHITE BRANT.
Rather rare. It arrives in Novem-
ber, and again late in February.
(S. & A.)

Blue-winged Goose. . *Anser cœrulescens.* This species, supposed by many authors to be the young of the preceding, *Anser hyperboreus,* with which it is generally found in company, is also rare, but in some seasons is not uncommon on the Delaware and Atlantic coast. (S. & A.)

White-fronted Goose. . *Anser albifrons.* LAUGHING GOOSE. Rare. Its migrations are more towards the interior States. (W.)

Canada Goose. . . . *Bernicla Canadensis.* WILD GOOSE. Not uncommon from the end of September to the beginning of April. (W.)

Brent Goose. . . . . *Bernicla brenta.* BRANT. Abundant, appearing early in October, and again in the middle of May. (S. & A.)

Mallard. . . . . . *Anas boschas.* WILD DUCK. GREEN HEAD. Common, but much more frequent in Winter than at other seasons.

Black Duck. . . . . *Anas obscura.* DUSKY DUCK. Abundant. There is a partial migration to the north in Spring, the flocks returning in October.

Pin-tail Duck. . . . *Anas acuta.* SPRIG-TAIL. Plentiful. This species leaves in the middle of March for the north. (W.)

Green-winged Teal. . *Nettion Carolinensis.* Abundant, appearing in April, and again about the end of October. (S. & A.)

Blue-winged Teal. . . *Querquedula discors.* Common. Arrives middle of April, and again in September. (S. & A.)

Shoveller. . . . . . *Spatula clypeata.* SPOONBILL. Rather rare. It leaves for the north about the middle of April. (W.)

Gadwall. . . . . . . *Chaulelasmus streperus.* GREY DUCK. WELSH DRAKE. Rare. (S. & A.)

Wigeon. . . . . *Mareca Americana.* BALDPATE. AMERICAN WIGEON. Not uncommon, arriving in September, and again in April. A few remain during Winter. (S. & A.)

Summer Duck. . . . *Aix sponsa.* WOOD DUCK. Plentiful. This beautiful species arrives early in April. (S.)

Scaup Duck. . . . . *Fuligula marila.* GREATER BLACK HEAD. BROAD BILL. BLUE BILL. Frequent, appearing early in October, and leaving in the end of April. (W.)

Lesser Scaup Duck. . *Fuligula affinis.* LITTLE BLACK HEAD. CREEK BROAD BILL. Not uncommon. It leaves for the north early in April. (W.)

Ring-necked Duck. . . *Fuligula collaris.* TUFTED DUCK. Frequent in Autumn, and appearing again late in March. (S. & A.)

Red-headed Duck. . . *Fuligula Americana.* POCHARD. Not uncommon, arriving from the north early in November. It is distinct from *Fuligula ferina* of Europe. (W.)

Canvas-back Duck. . . *Fuligula vallisneria.* Abundant, arriving early in November, and remaining for the Winter on the Susquehanna and Chesapeake Bay. (W.)

Golden Eye. . . . . *Fuligula clangula.* WHISTLER. WHISTLE WING. Plentiful from November to April. It may be a variety of the European bird, but the difference is very slight. (W.)

Buffel-headed Duck. . *Fuligula albeola.* BUTTER BALL. SPIRIT DUCK. Abundant from October to beginning of May. (W.)

Long-tailed Duck. . . *Harelda glacialis*. SOUTH SOUTHERLY. OLD SQUAW. Frequent during Winter along the river Delaware and the sea-coast. (W.)

Labrador Duck. . . *Camptolæmus Labradorius*. SAND-SHOAL DUCK. PIED DUCK. Rare. A few are seen every season. (W.)

Velvet Scoter. . . . *Oidemia fusca*. WHITE - WINGED COOT. Plentiful on the sea-coast. It arrives in the middle of October, and departs early in April. (W.)

Surf Scoter. . *Oidemia perspicillata*. BLACK SEA DUCK. Common from October to May along the Jersey coast. (W.)

Black Scoter. . . . *Oidemia Americana*. BROAD-BILLED COOT. Not uncommon. It resembles so closely the *Oidemia nigra* of Europe, as to be only a variety. (W.)

Ruddy Duck. . . *Erismatura rubida*. SPINE - TAIL. SALT-WATER TEAL. Rare. It is more abundant in the interior. (W.)

Goosander. . . . . *Mergus merganser*. SHELDRAKE. FISHER DUCK. DUN DIVER. Abundant from the beginning of November to April, but many breed in the interior and are resident. The bill of European specimens is somewhat longer and more slender.

Red-breasted Merganser. *Mergus serrator*. PIED SHELDRAKE. SAWBILL. Not uncommon. A few remain to breed. (W.)

Hooded Merganser. . . *Mergus cucullatus*. WATER PHEASANT. HAIRY HEAD. This handsome species is plentiful.

Gannet. . . . . *Sula bassana*. SOLAN GOOSE. Very rare. It is seen along the sea-coast more frequently in Winter than at other seasons. (W.)

F

Common Cormorant. . *Phalacrocorax carbo.* Rather rare
on the Jersey coast. (W.)

Double-crested Cormorant. *Phalacrocorax dilophus.* Rare;
many, however, pass along the coast
to Winter further south. (W.)

Wilson's Petrel. . . . *Thalassidroma Wilsonii.* Very rare.
It is generally met with off the coast
in Autumn and Winter. (W.)

Greater Shearwater. . *Puffinus major.* Very rare. A few
are seen every year on the Atlantic,
off the coast. (W.)

Great Black-backed Gull. *Larus marinus.* SADDLE BACK GULL.
Not uncommon. (W.)

Herring Gull. . . . *Larus argentatus.* GREY WINTER
GULL. SILVERY GULL. Plen-
tiful. (W.)

Ring-billed Gull. . *Larus Delawarensis.* COMMON GULL.
BROWN WINTER GULL. Abundant.
This species is not unlike *Larus
canus* of Europe. (W.)

Laughing Gull. . . . *Larus atricilla.* BLACK - HEADED
GULL. Common, arriving in the
end of April. (S.)

Bonaparte's Gull. . . *Larus Philadelphicus.* LESSER
BLACK-HEADED GULL. Not un-
common. (W.)

Kittiwake Gull. . . . *Larus tridactylus.* Rather rare along
the New Jersey coast. (W.)

Marsh Tern. . . . . *Sterna anglica.* GULL-BILLED TERN.
Rare. (S.)

Forster's Tern. . . . *Sterna Forsteri.* Rare. I have found
this species breeding on Brigantine
Beach. The adult, in Winter
plumage, has been called HAVELL'S
TERN. (S.)

Roseate Tern. . . . *Sterna Dougallii.* This elegant Tern
is not uncommon. It is doubtful if it
is increasing in Scotland, where the
species was originally discovered. (S.)

Common Tern. . . . *Sterna hirundo.* SUMMER GULL. Plentiful, arriving in the middle of April. (S.)

Arctic Tern. . . *Sterna arctica.* Rare. This species appears to be most numerous in Autumn, an occasional straggler only being observed in Winter. It breeds from New England northwards. (S.)

Least Tern. . *Sterna frenata.* LITTLE SHEEPSHEAD GULL. Common, arriving in May, and departing in August. It is rather larger than *Sterna minuta* of Europe, and is otherwise distinct. (S.)

Black Tern. *Sterna fissipes.* SHORT-TAILED TERN. Not uncommon. It breeds in the Western States, during which season it is absent from the coast. (S. & A.)

Black Skimmer. . . . *Rhynchops nigra.* SHEARWATER. CUTWATER. Frequent, arriving in the middle of May. (S.)

Great Northern Diver. *Colymbus glacialis.* LOON. Abundant; and although it is more frequent in Winter, many are found in Summer during the breeding season.

Red-throated Diver. . *Colymbus septentrionalis.* Rather rare; the specimens procured being mostly young birds. (W.)

Red-necked Grebe. . . *Podiceps rubricollis.* Rarely met with so far to the south. (W.)

Great-crested Grebe. . *Podiceps cristatus.* DIPPER. Not uncommon. (W.)

Sclavonian Grebe. . . *Podiceps cornutus.* HORNED GREBE. WATER WITCH. HELL DIVER. Frequent in Winter. A few remain to breed. (W.)

Pied-Bill Dobchick. . . *Podiceps Carolinensis.* DIPPER. CAROLINA GREBE. Common, arriving early in April. (S.)

Razor Bill. . . . . . *Alca torda.* RAZOR - BILLED AUK.
Very rare, coming early in Novem-
ber. A few migrate as far south as
Cape May. (W.)

Puffin. . . . . . . *Fratercula arctica.* SEA PARROT.
ARCTIC PUFFIN. An extremely
rare Winter visitant along the
coast. (W.)

Thick-billed Guillemot. *Uria Brunnichii.* BRUNNICH'S GUIL-
LEMOT. Very rare, but nearly every
Winter specimens are procured from
the Jersey sea-coast. (W.)

STRAGGLERS, OR IRREGULAR VISITANTS.

Swallow-tailed Kite. . *Nauclerus furcatus.* Has been seen
once or twice in Pennsylvania. Mr
John Krider shot one near Philadelphia in 1857. (S.)

Great Grey Owl. *Syrnium cinereum.* LAP OWL.
Although a northern species, this
Owl has been several times found
in New Jersey. (W.)

Hawk Owl. . . . . *Surnia ulula.* CANADA OWL. DAY OWL. Is occasionally found in severe Winters. One was shot at Haddington, near Philadelphia, in 1866. (W.)

Carolina Parrot. . . . *Conorus Carolinensis.* PARAKEET. Occurs at rare intervals in Southern Pennsylvania. (S.)

Savanna Blackbird. . *Crotophaga ani.* A specimen killed near Philadelphia, in September, 1849, is now in the museum of the Academy of Natural Sciences. (S.)

Arctic Three-toed Wood- *Picoides arcticus.* BLACK - BACKED
pecker. THREE-TOED WOODPECKER. This species is occasionally seen in the northern counties of Pennsylvania. Audubon met with it in the forests of the Pocono Mountains. It is distinct from the *Picoides tridactylus* of Europe.

Fork-tailed Flycatcher. *Milvulus tyrannus.* Bonaparte procured a specimen of this bird near Bridgeton, New Jersey; another was shot by Audubon at Camden, near Philadelphia, in June, 1832. (S.)

Arkansas Flycatcher. . *Tyrannus verticalis.* A specimen, shot at Moorestown, New Jersey, was obtained by Mr John Cassin in the flesh, and is in the Academy Museum, Philadelphia. (S.)

Varied Thrush. . . . *Turdus nœvius.* This inhabitant of the Pacific coast has once been procured in New Jersey.

Cuvier's Golden-crested *Regulus Cuvieri.* Audubon killed a
Wren. specimen on the banks of the river Schuylkill in June, 1812. It has not been seen since, and is only known from his description and figure.

Prothonotary Warbler.   *Protonotaria citrea.* Is an occasional straggler to the lower portions of Pennsylvania. Mr John Cassin shot one near Wilmington. (S.)

Orange-crowned Warbler. *Helminthophaga celata.* An example of this species was shot on Rancocas Creek early in February, 1860. Chris. Wood also procured a fine male in Bucks County, on 2d November, 1867.

Townsend's Warbler. . *Dendroica Townsendii.* A full-plumaged male was shot in Chester County, near the Brandywine, 12th May, 1868, and now enriches my collection. (S.)

Yellow-throated Warbler. *Dendroica superciliosa.* A southern species, and rare straggler to the lower counties of Pennsylvania and New Jersey. (S.)

Small-headed Flycatcher. *Myiodioctes minutus.* Wilson procured this bird in April and June in New Jersey. Audubon mentions having seen it in Kentucky, but it has not been found since, and is a doubtful species.

Bohemian Waxwing. . *Ampelis garrulus.* Has been occasionally shot near Philadelphia. It is not uncommon on Lake Superior. (W.)

Bartram's Vireo. . . . *Vireo virescens.* Professor Baird mentions the probability of this species being an occasional visitor. It is more properly an inhabitant of South America.

Brown-headed Nuthatch. *Sitta pusilla.* A southern species, and rare visitant to the lower counties. (S.)

Lapland Long-Spur. . *Plectrophanes Lapponicus.* LAPLAND LARK BUNTING. Very rare, and found only in severe Winters. (W.)

Townsend's Bunting. . *Euspiza Townsendii.* There is but one specimen of this bird known; it was procured in Chester County. This so-called species may be only a variety of *Euspiza Americana.*

Blue Grosbeak. . . . *Guiraca cœrulea.* A rare straggler to the southern counties of Pennsylvania and New Jersey. It arrives in the middle of May. (S.)

Yellow-headed Troopial. *Xanthocephalus icterocephalus.* YELLOW-HEADED BLACKBIRD. Dr Jackson mentions that this species is occasionally seen along the Alleghany mountains, where a flock appeared in the Autumn of 1857. Mr John Krider shot a young male near Philadelphia. (S.)

Canada Jay. . . . . *Perisoreus Canadensis.* WHISKY JACK. A rare straggler to the northern counties of Pennsylvania. (W.)

Ground Dove. . *Chamaepelia passerina.* Mr John Krider shot a specimen near Camden in the Autumn of 1858. According to Dr Barton, who gave a list of the Birds of Pennsylvania in 1799, it was an occasional visitant to the neighbourhood of Philadelphia about a century ago. (S).

Whooping Crane. . . *Grus Americanus.* STORK. WHITE CRANE. Now very rare. While at Beasley's Point in 1857, I saw three off the inlet; they were very wary and could not be approached. In Wilson's time it bred at Cape May.

Louisiana Heron. . . *Ardea Ludoviciana.* This species has occasionally been obtained on the New Jersey coast. (S.)

Yellow-crowned Heron. *Nyctherodius violaceus.* A rare straggler from the south. It has been seen on the borders of the Schuylkill, near Philadelphia. (S.)

White Ibis. . *Ibis alba.* WHITE CURLEW. A very rare visitant so far north. I shot one at Great Egg Harbour in the Summer of 1858. (S.)

Glossy Ibis. . . *Ibis falcinellus.* Last season (1866) Mr John Krider shot a specimen just below Philadelphia. At long intervals, it has been seen on the river Delaware, and also at Egg Harbour. (S.)

Wilson's Phalarope. . *Phalaropus Wilsonii.* A very rare straggler. On the authority of Audubon, it has been stated that it breeds at Egg Harbour. (S. & A.)

Red Phalarope. . . . *Phalaropus fulicarius.* Called GREY PHALAROPE in Great Britain, where it is seldom found in Summer plumage. A few examples of this species are obtained every season; one was shot last September on the Delaware, at the mouth of Timber Creek, and shown to me by my friend, Mr B. A. Hoopes. (S. & A.)

European Woodcock. . *Scolopax rusticola.* A specimen was killed at Shrewsbury, New Jersey, in 1859.

Curlew Sandpiper. . . *Tringa subarquata.* Occasionally shot at Egg Harbour. Wilson must have met with it, as in his portfolio of drawings I found a figure of this bird in Autumn plumage. (S. & A.)

Ruff. . . . . . *Machetes pugnax.* Stragglers from Europe have occasionally been found on Long Island, and a single individual was met with on the coast of New Jersey.

Corn Crake. . . *Crex pratensis.* A specimen shot at Salem is now in the collection of the Academy of Sciences. Another was procured near Bordentown, New Jersey, by Mr John Krider. It is known as a Summer visitant to Greenland. (S.)

Purple Gallinule. *Gallinula martinica.* A very rare straggler from the south; it has been met with on the Jersey coast; and Mr John Krider informs me that he shot a fine specimen on League Island in September, 1848. (S.)

Trumpeter Swan. . . *Cygnus buccinator.* This noble bird —peculiar to the Continent west of the Mississippi—is included, on the authority of reliable sportsmen, who have shot it on the Chesapeake, as also Delaware Bay. It must be a rare straggler, however, and I have not noticed it in the Philadelphia market, where the Wild or "Whistling Swan" (*Cygnus Americanus*) is so frequently seen every Winter. (W.)

Lesser Snow Goose. . *Anser albatus.* Mr John Cassin procured, in the Philadelphia market, two pairs, in the course of twenty years, of this inhabitant of North-West America. (W.)

Hutchin's Goose. *Bernicla Hutchinsii.* Occasionally seen as far south as Chesapeake Bay. (W.)

G

Black Brent Goose. . . *Bernicla nigricans.* BLACK BRANT.
One was procured by Mr George
N. Lawrence at Egg Harbour in
January, 1846; and flocks have
been seen since at rare intervals.
Mr John Krider acquaints me of
having received several specimens
from the coast. (W.)

English Teal. . . . *Nettion crecca.* GREEN-WINGED TEAL.
Has occasionally been met with.
Mr J. G. Bell, New York, tells me
it is not very uncommon on the
shores of Nova Scotia and New-
foundland. (W.)

English Wigeon. . *Mareca penelope.* WIGEON. This
is also a straggler from Europe,
some being met with almost every
season. (W.)

Harlequin Duck. . . *Fuligula histrionica.* LORD AND LADY
DUCK. A very rare visitant from
the north to the sea-shore. (W.)

Eider Duck. . . . . *Somateria mollissima.* SHOAL DUCK.
Has been seen occasionally at Egg
Harbour. (W.)

King Duck. . . . . *Somateria spectabilis.* KING EIDER.
Has also been observed at Egg Har-
bour during severe Winters; the
specimens obtained being generally
young birds. (W.)

Rough-billed Pelican. . *Pelecanus erythrorhynchus.* WHITE
PELICAN. Has been seen at rare
intervals on the Delaware, and on
the sea-coast near Cape May.

Brown Pelican. . . *Pelecanus fuscus.* One was shot off
Sandy Hook in 1837. (S.)

Leach's Petrel. . . . *Thalassidroma Leachii.* FORK-TAILED
PETREL. Occasionally seen off the
coast. During a gale in August,
1842, a number were driven inland.

Stormy Petrel. . . . *Thalassidroma pelagica.* Mother Carey's Chicken. Has been found off the coast. One was captured under Market Street Bridge a few years ago.

Manx Shearwater. . . *Puffinus anglorum.* An accidental visitor to the coast, generally appearing in Autumn. (S.)

Dusky Shearwater. . . *Puffinus obscurus.* An occasional straggler along the sea-shore, from the south. (S.)

Pomarine Skua. . . . *Stercorarius pomarinus.* Has been found on the coast in Winter. One specimen was taken at Harrisburg in Summer. (W.)

Arctic Skua. . . . . *Stercorarius parasiticus.* Richardson's Skua. Sometimes appears on the coast. Mr Krider shot one on the meadows to the south of Philadelphia. (W.)

Caspian Tern. . . . *Sterna Caspia.* Specimens have been procured from the coast of New Jersey at rare intervals. (W.)

Royal Tern. . . . . *Sterna regia.* Also very rare. This bird is distinct from *Sterna Cayana* of South America. (S.)

Sandwich Tern. . . . *Sterna cantiaca.* A specimen of this straggler from the Gulf States was shot on Grassy Bay in August, 1861. (S.)

Black Guillemot. . . *Uria grylle.* Guillemot. Occasionally migrates as far south as Cape May in Winter. (W.)

Bridled Guillemot. . *Uria lachrymans.* Murre. Ringed Guillemot. Foolish Guillemot. Occasionally met with on the coast in Winter. It is doubtful if the true *Uria troile* of Europe has yet been seen in the United States.

For an account of the distribution
of this bird in the British Islands,
see "Gray's Birds of the West of
Scotland." (W.)

Little Auk.   .   .   .   .   *Mergulus alle.*   LITTLE ROTCHIE.
Occasionally shot at Egg Harbour
and along the coast. (W.)

BIRDS WHICH HAVE DISAPPEARED.

CONCLUDING REMARKS.

SINCE the eastern provinces have become more densely populated,
many of the larger and more wary species of birds have changed
their course of migration, and now reach the arctic regions by a
route taking them towards the interior of the continent; and there
are also some, formerly known as Summer visitants, which have now
a more southern limit.   Parrots, for example, (*Conorus Carolinensis*), are at the present day rarely found north of the Carolinas;
while Wild Turkeys, which were once abundant, although still
to be met with in suitable localities, are now in very limited

numbers. In a rare tract printed in 1648, entitled "A Description of New Albion"—a name at one time applied to this part of the country—we read of four or five hundred Turkeys forming a single flock. The Pinnated Grouse (*Cupidonia cupido*) is another interesting bird which has become nearly extinct in East Pennsylvania, and entirely so, it is believed, in New Jersey. The Whooping Crane (*Grus Americanus*) may also be said to have disappeared, not even a straggler having been seen for some years. It likewise seems to have been once very plentiful; for we read in Hakluyt's Voyages, Ed. 1589, fol. 729, that Captain Philip Amadas and his fellow-adventurers, who visited and explored the coast in the year 1584, "having discharged their harquebus-shot, such a flocke of Cranes (the most part white) arose, with such a crye, re-doubled by many ecchoes, as if an armie of men had showted altogether." The Brown or Sandhill Crane (*Grus Canadensis*) has not been seen in this region for many years past, although it is still not uncommon in the west. The learned Professor Kalm, who travelled in this country in 1748–49, and resided some time at Swedesborough, N.J., noticed this bird on its northern flight about the middle of February. At that time they usually alighted, but remained for a short time only, every Spring, in comparatively limited numbers; but he was assured by a colonist, above ninety years of age, that in his youth (or about the year 1670) Cranes came in hundreds. The Rough-billed Pelican (*Pelecanus erythrorhynchus*) was also frequent on the Hudson and the Delaware, but is now a very rare visitant to the last-mentioned river only.

While ornithologists, however, have to deplore the diminution in the number of the more conspicuous birds which has taken place during the last century, it is gratifying to find a very sensible increase in the number of other species. Many of the Warblers, for example, then considered rare, are now found to be abundant— a beneficial increase, for which we are no doubt indebted to the fact of our efficient game laws providing for the protection of insectivorous birds. But on the other hand, the constant shooting of "Bay Snipe" and shore birds generally, by market gunners, always on the watch for their arrival, has seriously reduced the flocks of many species formerly known to abound in districts now but thinly peopled by this interesting class. The late Mr George

Ord assured the writer that, during his excursions to the coast with Wilson, the distinguished ornithologist, the Avocet, Stilt, and other Waders which are becoming rare in our day, were then quite plentiful; so that there is every reason to fear that, in the course of a few years more, they also may disappear. In Chesapeake Bay, the Winter resort of a great variety of wild fowl, birds, although still numerous, are, through the same influences, becoming every year less abundant; and unless the present reprehensible and most destructive system of shooting—wholesale slaughter, it may with propriety be called—be rigidly put down, the decrease will, in all likelihood, become permanent, to the great regret of every true-minded naturalist. We have but to look into the history of some of the birds of the British Islands, as a warning against the continuance of the destroying influences to which many of those of our own country are now subjected. During the past thirty years the Rapacious Birds of Great Britain have undergone an amount of persecution so determined and systematic, that many of the species have altogether disappeared; and as by the latest records of the meetings of the British Association assembled at Norwich, it would appear that even sea-fowl are now in danger of extirpation, owing to the extraordinary demand for their plumes and feathers for marketable purposes, it may not be out of place for the ornithologists of this great continent to consider the propriety of protecting, even now, some of the species thus proclaiming by their scarcity, that the time may not be far distant when we too may have to lament their loss.

THE END.

GLASGOW: PRINTED BY ARCH. K. MURRAY AND CO.